Questions in
Intermediate 2
Physics

**Richard Bush and
Neil Short**

Leckie × Leckie
Scotland's leading educational publishers

Text © 2004 Richard Bush and Neil Short

Design and layout © 2004 Leckie & Leckie

Cover image © Photodisc

03/131207

Published by
Leckie & Leckie
3rd Floor, 4 Queen Street
Edinburgh
EH2 1JE

Tel: 0131 220 6831
Fax: 0131 225 9987
Email: enquiries@leckieandleckie.co.uk
Web: www.leckieandleckie.co.uk

Special thanks to: River Design (cover design, page layout and illustrations) and Bruce Ryan (project management)

ISBN 978-1-84372-149-9

A CIP Catalogue record for this book is available from the British Library.

® Leckie & Leckie is a registered trademark.

Leckie & Leckie is a division of Huveaux plc.

Contents

Introduction

This book has been designed to help you pass your Intermediate 2 Physics examination by giving you the opportunity to practice questions covering all four units of the Intermediate 2 course.

The questions are in two types: feeders and exam-style. The feeder questions are shorter ones which relate to the same work as the exam-style question which follows. Exam-style questions are marked with this icon: We recommend that before tackling an exam-style question, you try the feeders associated with it as a way of warming yourself up.

In the final examination you will not be allowed to look up equations and prefixes. You should therefore be aiming to be able to do the questions in this book without having to look anything up! However, the *Useful Physics Data* section at the back of this book lists the most important equations and prefixes should your memory need a little help.

We have included a section with the answers to all the questions. For the feeder questions we have given only numerical answers with units. For the longer main questions we have given more detailed solutions so that you can learn how they are done should you find them a little difficult.

We hope you find this book useful and wish you good luck with your Intermediate 2 Physics.

Richard Bush   Neil Short

2 Mechanics and Heat

Scalars and Vectors

1

    a Bunny covers 200 m in 50 s. Find her average speed. **b** How far does the car travel in 60 s? **c** How long would it take the rocket to travel 80,000 m?

2 Calculate the missing numbers in this athletics recording sheet using the appropriate fomula:

Runner	Race Distance	Time	Average Speed
B. Bloggs	100 m	12·5 s	
H. Gebresellasie	5000 m		6·25 m/s
S. Motion		8 min 20 s	3 m/s

3 Look at these car speedometers.

    a   **b**   **c**

    Find the distance covered by the cars in 30 minutes. Write your answer in metres. (1 mile = 1600 m.)

4 How long would it take these moving objects to travel 1 kilometre?

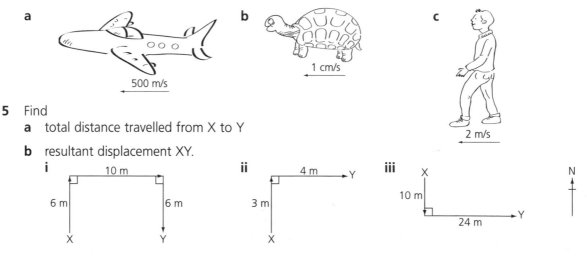

    a   **b**   **c**

5 Find
    a total distance travelled from X to Y

    b resultant displacement XY.

    i 10 m / 6 m / 6 m / X / Y   **ii** 4 m / Y / 3 m / X   **iii** X / 10 m / 24 m / Y / N

6 Each journey in **5** above was completed in 10 s. Find the average speeds and average velocities.

7 Use the card lengths and the readings on the timers to find the instantaneous speeds of the two trolleys:

    a 10 cm `0.20s`   **b** 5 cm `0.05s`

Michael builds a maze to test his two pet rats Dozey and Einstein. The diagram below shows a view from above the maze

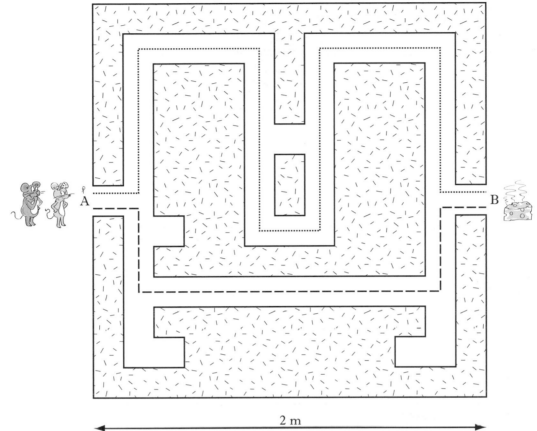

2 m

He starts the rats off at A and waits until they reach the exit at B. Dozey takes 11 seconds to follow the path shown by the dotted line. He travels 5·5 m. Einstein takes 6 seconds to follow the path shown by the dashed line. Einstein and Dozey run with exactly the same average speed.

Use the information to find:

a Dozey's average speed

b The distance covered by Einstein

c The final displacement of each rat

d Dozey's average velocity

e Einstein's average velocity.

Projectiles

1 This ball is travelling along a smooth horizontal surface at 12 m/s.

How far will it travel in

a 1·5 s

b 3·0 s?

2 The acceleration due to gravity is approximately 10 m/s². These balls were released from rest.

    a   **b**

           t = 1·5 s   t = 3·0 s

How fast is each ball travelling at the times shown?

3 This ball is projected horizontally at V off the edge at a speed of 12 m/s and eventually hits the ground at X after 3 seconds.

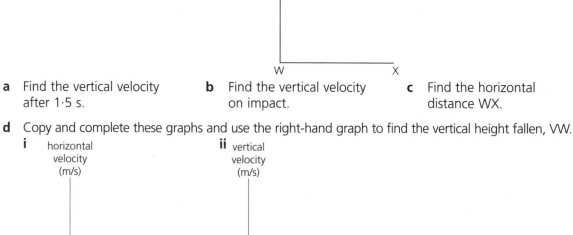

    a Find the vertical velocity after 1·5 s.   **b** Find the vertical velocity on impact.   **c** Find the horizontal distance WX.

    d Copy and complete these graphs and use the right-hand graph to find the vertical height fallen, VW.

       i horizontal velocity (m/s)   **ii** vertical velocity (m/s)

               0   1 2 3 t (s)   0   1 2 3 t (s)

    e Using information from the above graphs, calculate the resultant velocity on impact at X.

See also Leckie & Leckie's Intermediate 2 Physics Course Notes

A pupil makes up a model bomber for a school fair.

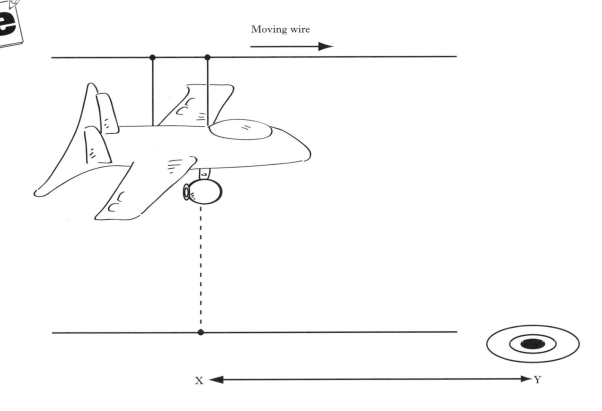

Moving wire

X ← → Y

The bomber moves horizontally at a speed of 1·2 m/s. The 'bomb' is released when it is above point X. It hits the centre of the target at Y, 0·5 seconds later. Neglect air resistance.

a Draw a speed/time graph of the horizontal motion of the bomb.

b Find the horizontal distance from X to Y.

c The vertical acceleration of the bomb is 10 m/s². Draw a speed/time graph of the vertical motion of the bomb.

d Use the graph to find the vertical height fallen.

e The pupil increases the horizontal speed of the bomber to 2 m/s and the bomb is still released above point X. How far beyond Y will the bomb now land?

See Answers on page 1 of answer booklet

Graphs of motion

1 Find the accelerations from these velocity/time graphs.

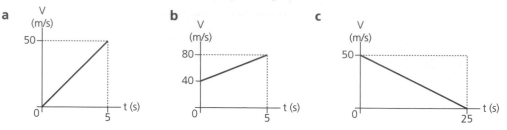

a

b

c

2 Find the distances travelled from these velocity/time graphs.

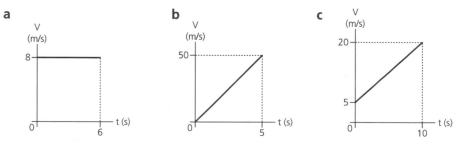

a

b

c

3 A car, initially at rest at traffic lights, accelerates to 15 m/s in 10 s, continues at that speed for a further 10 s and then brakes to a halt in 3 s.

    a Use the above information to draw a velocity/time graph of the journey.

    From the graph, find:

    b the acceleration during the first 10 s

    c the deceleration during the final 3 s

    d the total distance travelled

    e the average speed of the car.

4 Use this graph to answer the following questions.

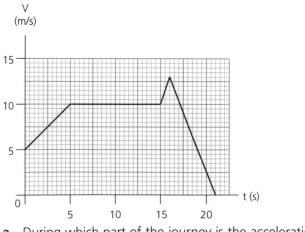

    a During which part of the journey is the acceleration greatest?

    b What is the total distance travelled?

    c What is the average speed for the whole journey?

See also Leckie & Leckie's Intermediate 2 Physics Course Notes

Courageous Charlotte is a competitor in a ski jumping competition. Each competitor skis down a runway before launching off the end of a ramp as shown in the diagram below.

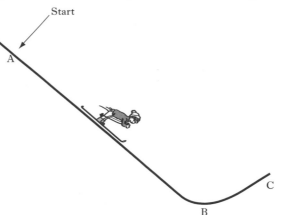

During one attempt, Charlotte's speed was recorded as she travelled down the runway and take-off ramp and a speed-time graph was plotted.

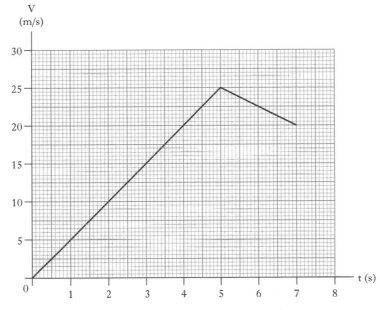

a   Using the graph, calculate Charlotte's acceleration during the downhill section of her run.

b   Calculate the distance she travelled between starting her run and reaching point B on the runway.

c   What was Charlotte's take-off speed at C?

d   Calculate the distance from B to C

e   Calculate Charlotte's average speed as she travels from A to C.

Forces

1 a Find Fred's weight.

mass = 60 kg
g = 10 N/kg

b Find Robot's mass.

weight = 60 N
g = 1·6 N/kg

c Find the gravitational field strength.

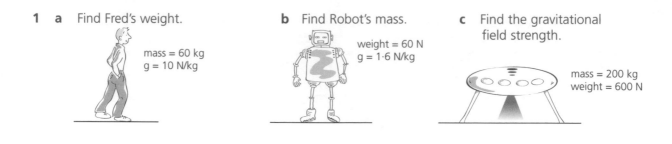

mass = 200 kg
weight = 600 N

2 Calculate the missing numbers in this table, using the appropriate formula.

Weight (N)	Mass (kg)	Gravitational Field Strength (N/kg)
100	5·0	
140		7·0
200	0·8	
	50·0	8·0
32		16·0
	75·0	1·2

3 Calculate the mass of a packet of dried 'space food' that weighs 80 mN on the Moon. The gravitational field strength is 1·6N/kg.

4 Calculate the weight of 300 g of water on a planet with gravitational field strength 13 N/kg.

5 a What forward force does Lance produce?

constant speed

air resistance
50 N

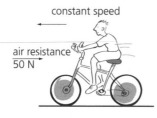

b What upward force do the wings produce?

horizontal flight

weight = 8N

6 a Calculate the acceleration of the wheelbarrow.

unbalanced force 90 N

mass 45 kg

b Calculate the mass of the car.

unbalanced force 500 N
acceleration = 0·4 m/s^2

c Calculate the unbalanced force on the bus.

mass = 16,000 kg
acceleration = 0·25 m/s^2

See also **Leckie & Leckie's Intermediate 2 Physics Course Notes**

7   Calculate the missing numbers in this table using the appropriate formula.

Force (N)	Mass (kg)	Acceleration (m/s²)
	50	2
45		15
100	25	
	300	10
270		0·3
4	20	

8   A force of 50 mN acts on a toy car, which has a mass of 80 g. Calculate the acceleration of the car.

9   If a force of 5 kN produces an acceleration of 40 cm/s², what is the mass of the object being accelerated?

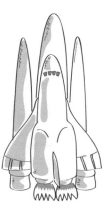

The following table gives some information about the space shuttle.

Total mass at take-off(kg)	Thrust of each main engine at take-off (N)	Thrust of each solid rocket booster at take-off (N)	Mass of shuttle in orbit (kg)	Maximum thrust of each main engine in orbit (N)
2,000,000	1,700,000	12,000,000	75,000	3,000

a   What is the weight of the shuttle at take-off? Assume g = 10 N/kg.

b   The shuttle has three main engines and two solid rocket boosters. Find the total thrust available at take-off.

c   Find the resultant force on the shuttle at take-off.

d   Find the acceleration at take-off.

e   Once the shuttle is in orbit, it can manoeuvre using main engine thrust. Find the maximum acceleration that can be produced by the main engines in orbit.

f   On re-entry to the earth's atmosphere, the shuttle's speed reduces dramatically without any use of the engines. What force is responsible for the deceleration?

See Answers on page 2 of answer booklet

2 Mechanics and Heat

Momentum and collisions

1 a Calculate the momentum of this athlete.

mass 55kg
5 m/s

b Calculate the mass of this bird.

10 m/s

momentum = 15 kg m/s

c Calculate the speed of the ball.

mass = 0· 2 kg

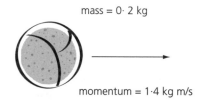

momentum = 1·4 kg m/s

2 Calculate the missing numbers in this table using the appropriate formula.

Momentum (kg m/s)	Mass (kg)	Velocity (m/s)
	90	3
500		25
240	60	
	1100	15
3000		1·5
120	12	

3 Calculate the momentum of a 15 g bullet which is fired with a velocity of 0·4 km/s.

See also Leckie & Leckie's Intermediate 2 Physics Course Notes

a A policeman uses a handheld radar gun to measure the speed of a moving vehicle. The gun sends out a pulse of microwaves and detects the pulse reflected back from the vehicle. The pulse travels at 3×10^8 m/s.

The policeman fires the radar at a speeding sports car. The reflected pulse is received one microsecond later. Calculate the distance between the car and the policeman.

b The sports car driver sees the policeman and briefly brakes. A 3000 kg lorry, travelling at 30 m/s behind the 1000 kg sports car, crashes into the back of the sports car. The vehicles lock together and continue on at 25 m/s.

  i Calculate the total momentum of the vehicles just after the collision.

  ii What was the total momentum of the vehicles just before the collision?

  iii Calculate the speed of the sports car just before the collision.

See Answers on page 3 of answer booklet

Energy

1 **a** Calculate the work done in pushing the trolley.

20 N
distance = 80 m

b How far does the person drag the sledge?

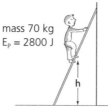
force = 6 N
work done = 300 J

c Calculate the force Elton produces when pushing his piano.

work done = 800 J
distance = 2 m

2 Calculate the missing numbers in this table using the appropriate formula.

Work Done (J)	Force (N)	Distance (m)
	50	3
300		75
4500	30	
	150	1200
3000		15
120000	600	

3 **a** Calculate the potential energy of the piano.

mass 300 kg
9 m

b Calculate the height, h, climbed by Hugh.

mass 70 kg
E_P = 2800 J
h

c Calculate the mass of the schoolbag.

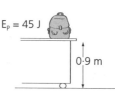
E_P = 45 J
0·9 m

4 Calculate the missing numbers in this table using the appropriate formula. Assume g = 10 N/kg.

Potential Energy (J)	Mass (kg)	Height (m)
	15	3·2
150		12
250	10	
	0·75	2·8
730,400		913
234,000	1300	

5 If a 1500 kg helicopter has a potential energy of 1·35 MJ, what is its height?

6 A 10 g mass is placed on a bench and has 80 mJ of potential energy. What is the height of the bench?

See also Leckie & Leckie's Intermediate 2 Physics Course Notes

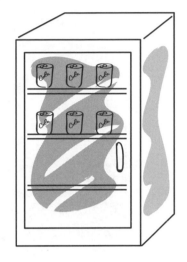

Sam keeps a small refrigerator in his bedroom to store cans of his favourite 'Superfizz Cola'.

Each can contains 300 g of cola. The specific heat capacity of the cola is 4000 J/kg°C.

He transfers 6 cans of cola in his bedroom at 20°C into the fridge which cools them down to 5°C.

a Find the total amount of heat energy the fridge must remove from the cola in all 6 cans to cool them from 20°C to 5°C.

b If the fridge removes heat from the cola at a rate of 100 joules per second, how long should the fridge take to cool all 6 cans down to 5°C ?

c Sam has an extra can which he wants to chill quickly. He takes it downstairs and places it in the kitchen freezer then unfortunately forgets about it! If the cola is initially at 20°C and it freezes at 0°C, find the minimum time for the cola to freeze solid at 0°C.
(Latent heat of fusion of cola = 300,000 J/kg. The freezer removes heat from the cola at a rate of 57 joules per second.)

2 Mechanics and Heat

Mechanical energy/heat

1 Both metal blocks were supplied with 80,000 J of energy. Calculate the specific heat capacities of the two metals X and Y.

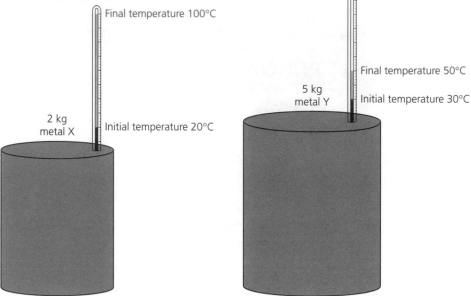

Final temperature 100°C

2 kg metal X
Initial temperature 20°C

5 kg metal Y

Final temperature 50°C
Initial temperature 30°C

2 Aluminium has a specific heat capacity of 880 J/kg°C. Find the temperature rise of each lump of aluminium if they each receive 10,560 J of energy.

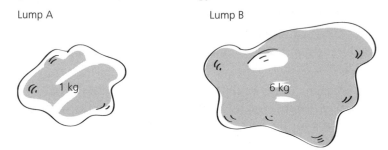

Lump A

1 kg

Lump B

6 kg

3 A 100 g lump of copper loses 19,000 J of heat when it is removed from an oven at 520°C and allowed to cool down to the temperature of the room. Find the room's temperature.
(Specific heat capacity of copper = 380 J/kg°C)

4 A lead bullet travelling at 260 m/s collides with a target and comes to rest. If all of its kinetic energy changes to heat, find the temperature rise of the bullet.
(Specific heat capacity of lead = 130 J/kg°C)

See also Leckie & Leckie's Intermediate 2 Physics Course Notes

Sarah is driving her sports car on a warm dry day in the summer. The total mass of the car with Sarah in it is 850 kg. On a straight section of road with good visibility Sarah's car reaches a speed of 30 m/s.

a   Calculate the total kinetic energy when travelling at 30 m/s.

b   Sarah sees a deer on the road so she brakes sharply and slows down to a speed of 12 m/s. How much kinetic energy is lost in slowing down from 30 m/s to 12 m/s?

c   The car has disc brakes that work when two stationary brake pads squeeze a rotating metal disc as shown in the diagram below.

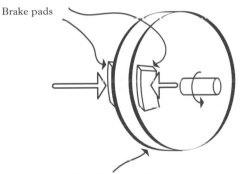

Brake pads

Rotating brake disc

What main energy transformation takes place when the brakes are operated to slow the car down?

The brake discs have a total mass of 7·6 kg. The specific heat capacity of the metal they are made from is 700 J/kg°C. Before the brakes are applied the discs are at a temperature of 25°C.

d   If all the kinetic energy lost in part **b** is transferred to the brake discs, calculate

   i   the temperature rise of the brake discs immediately after the car has slowed down

   ii   the final temperature of the brake discs immediately after the car has slowed down.

e   In practice the final temperature of the brake discs would not be as high as calculated in part **d ii**. Explain why.

Electrostatics

1 a How much charge passes through the lamp in 15 s?

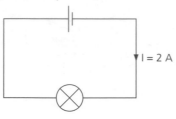

b How long does it take for 270 C of charge to pass through the resistor?

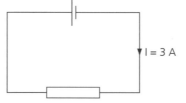

c 35 C of charge passes through the lamp in 70 s. How much current is flowing?

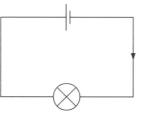

2 Calculate the missing numbers in this table using the appropriate formula.

Current	Charge	Time
10 A	45 C	
	180 C	60 s
2·3 C		20 s
0·25 A	1200 C	
	6 C	1·2 s
1·5 A		90 s

3 Calculate how much charge passes through an LED in 1 minute if the current is 20 mA.

4 Calculate the current drawn by a kettle if 3000 C of charge passes through it in 5 minutes.

After a long run in his sports car, Ralph has become negatively charged by friction between his clothes and the material of the car seat. The negative charge on him is 6×10^{-6} C.

a   Which particles are responsible for his negative charge?

b   Explain why the car seat should have become positively charged.

When he gets out of the car and touches the metal car door, the charge passes from him to the car and he feels a small electric shock. The shock lasts for a millisecond.

c   Calculate the current flowing from Ralph to the car.

Ralph's sports car has a battery with a capacity of 36 ampere hours. (This means that it can supply 36 amperes for one hour, 18 amperes for two hours, 9 amperes for 4 hours, etc.) Because the sports car is not driven regularly, its battery tends to become discharged. Ralph buys a pack of solar cells and connects them to the battery. In sunny conditions, the cells charge the battery with a current of 50 mA.

d   How many coulombs of charge can the battery store when it is fully charged?

e   Describe the energy change that takes place when the solar cells are charging the battery.

f   If the battery was completely discharged, show that it would take the solar cells 90 days to fully charge it, assuming 8 hours of sunshine each day.

3 Electricity and Electronics

Circuits – series/parallel

1 What are the readings on A_2 and V_2?

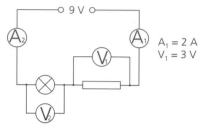

$A_1 = 2$ A
$V_1 = 3$ V

2 What is the reading on V_2 if the three lamps are identical?

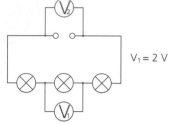

$V_1 = 2$ V

3 What is the supply voltage V_S?

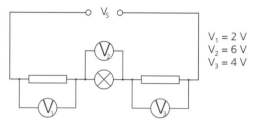

$V_1 = 2$ V
$V_2 = 6$ V
$V_3 = 4$ V

4 What are the readings on V_1 and A_3?

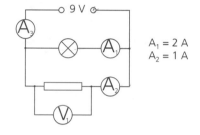

$A_1 = 2$ A
$A_2 = 1$ A

5 What is the reading on A_2 if the three lamps are identical?

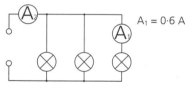

$A_1 = 0.6$ A

6 What is the supply voltage V_S and the reading on A_2?

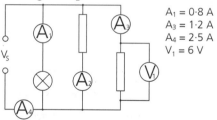

$A_1 = 0.8$ A
$A_3 = 1.2$ A
$A_4 = 2.5$ A
$V_1 = 6$ V

7 Calculate the resistance of the lamp.

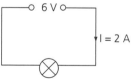

6 V
$I = 2$ A

8 Calculate the supply voltage, V_S.

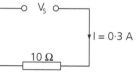

V_S
$I = 0.3$ A
10 Ω

9 Calculate the current, I.

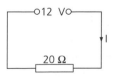

12 V
I
20 Ω

10 What are the total resistances of these resistor combinations?

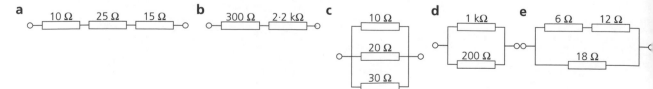

a 10 Ω 25 Ω 15 Ω
b 300 Ω 2·2 kΩ
c 10 Ω 20 Ω 30 Ω
d 1 kΩ 200 Ω
e 6 Ω 12 Ω 18 Ω

See also Leckie & Leckie's Intermediate 2 Physics Course Notes

11 Calculate the resistance of a light bulb if a current of 0·65 A is drawn when connected to the mains supply of 230 V.

12 Calculate the starting current if a car starter motor has a resistance of 0·24 Ω and is connected to a 12 V battery.

A physics laboratory assistant makes up resistors by wrapping resistance wire around wooden rods. Using wire rated at 20 ohms per metre he makes up two different types of resistor: type X and type Y.

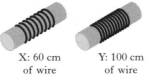

X: 60 cm
of wire

Y: 100 cm
of wire

Type X is made from 60 centimetres of wire and type Y from 100 centimetres of wire.

a Calculate the resistance of a type X resistor.

b Calculate the resistance of a type Y resistor.

Calculate the total resistance of each of these combinations of his resistors:

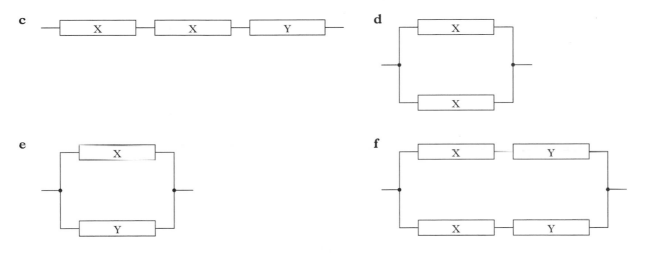

The resistors are then connected into the two circuits shown below. Find the readings on the ammeters and voltmeters.

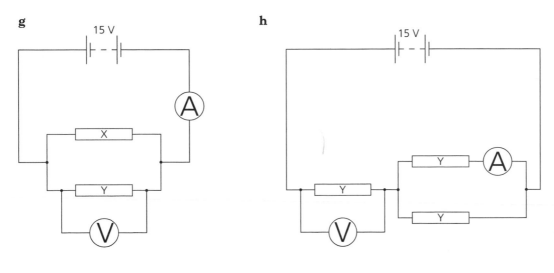

Circuits – power

1   Find the power dissipated in each circuit:

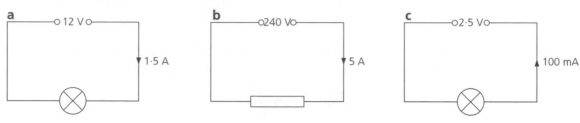

2   Find the currents:

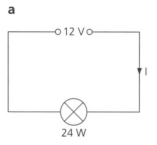   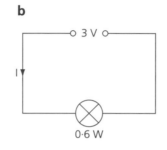   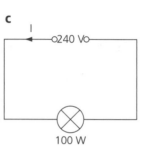

3   Find the supply voltages:

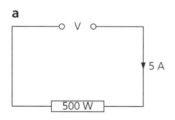   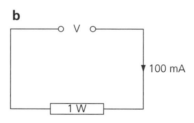   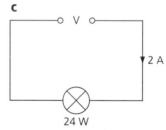

4   Find the resistances of the resistors when operating as shown:

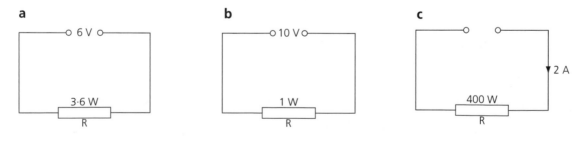

The diagram below shows a 6 V bicycle lamp circuit. A 3-position switch allows it to be
- Off
- On at normal power or
- On at reduced power.

The lamp is rated at 12 W.

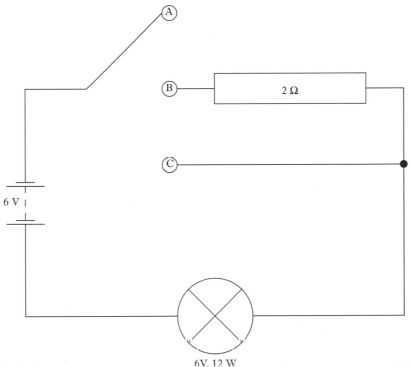

a In which position (A, B or C) will the lamp be off?

b In which position will the lamp operate at normal power?

c Find the current in the lamp when the switch is at position C.

d Find the resistance of the lamp when the switch is at position C.

e With the switch at position B, the current in the lamp is 1·5A. Assuming the resistance of the lamp does not change, find the new power rating of the lamp.

f State one advantage and one disadvantage of operating the bicycle lamp at reduced power.

Oscilloscope – AC/DC

1 a What is the peak voltage?

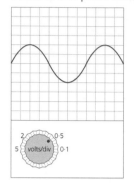

b If the peak voltage is 7·5 V, what is the setting on the volts/division control?

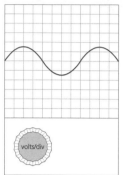

Alan is given a power supply a lamp and an oscilloscope by his physics teacher and asked to investigate the difference between two sets of terminals on the power supply. One pair of terminals is yellow and the other pair has one black terminal and one red one. Both sets of terminals are labelled '6 V'. Alan obtains oscilloscope traces as shown in the diagrams below.

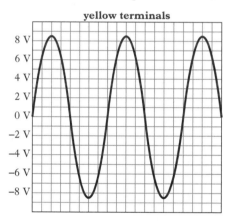

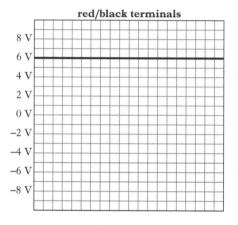

a Which set of terminals gives an a.c. output?

Alan connects the lamp to the yellow terminals.

b Describe the motion of the current in the lamp.

Alan connects the lamp to the red and black terminals expecting its brightness to be dimmer. In fact its brightness is the same.

c Explain why the lamp has the same brightness when connected to either set of terminals.

d When the yellow terminals were labelled 6 V, what voltage were the manufacturers referring to?

See also Leckie & Leckie's Intermediate 2 Physics Course Notes

Electromagnetism

1 Each coil below is connected to a d.c. electric supply.

  20 turns 1·5 A

  40 turns 1·5 A

  40 turns 3·0 A

Which coil should have

a the strongest magnetic field

b the weakest magnetic field?

2 A pupil has performed a series of experiments moving a bar magnet and different coils to generate voltages. The table below shows the results of his experiments. All the voltages have been measured correctly but unfortunately he has written them in the wrong order. Write them out in the correct order.

	Experiment	Induced voltage (V)
a	20 turns — N S 1 m/s, stationary	0
b	40 turns — N S stationary, ← 1 m/s	3·6
c	40 turns — N S 1 m/s →, ← 1 m/s	1·8
d	20 turns — N S 1 m/s →, 1 m/s →	5·4
e	60 turns — N S 1 m/s →, ← 1 m/s	0·9

The diagram below shows the inside of the pick-up from a record deck. The stylus runs in the groove on the record. Movement of the stylus in the record's groove causes a small magnet to move in front of the coil of wire.

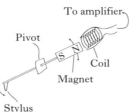

a Explain why there is a voltage output from the coil of wire when playing a record.

b Will this output be a.c. or d.c. ?

c Explain what you would expect to happen to the output voltage from the coil as the frequency of movement increases.

d What two changes could be made to the design of this pick-up to give a higher output voltage from the coil?

3 | Electricity and Electronics

Transformers

1 Find the voltmeter readings for these ideal transformers.

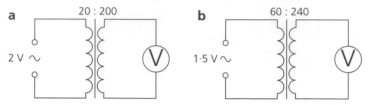

a 20 : 200 2 V ∿

b 60 : 240 1·5 V ∿

c 150 : 30 10 V ∿

2 Find the ammeter readings for these ideal transformers.

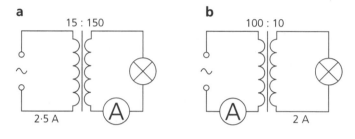

a 15 : 150 2·5 A

b 100 : 10 2 A

3 Find the currents I$_P$ and I$_S$ for these ideal transformers.

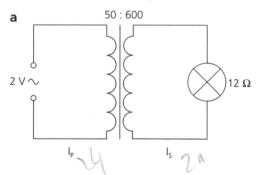

a 50 : 600 2 V ∿ 12 Ω I$_P$ I$_S$

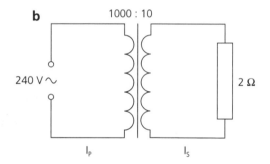

b 1000 : 10 240 V ∿ 2 Ω I$_P$ I$_S$

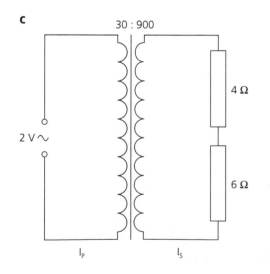

c 30 : 900 2 V ∿ 4 Ω 6 Ω I$_P$ I$_S$

See also Leckie & Leckie's Intermediate 2 Physics Course Notes

On a building site Terry uses power tools that operate at 110 V a.c. rather than the usual UK mains supply of 230 V a.c. The power tools are connected to a transformer which is plugged into the mains electrical supply.

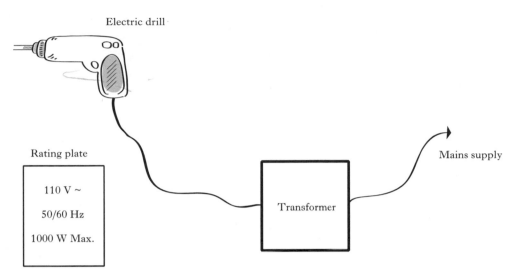

The rating plate of an electric drill that Terry is using states 1000 W, 110 V.

a Calculate the current that is drawn by the drill when operating at full power.

b Assuming that the transformer is an ideal transformer, calculate the current drawn from the mains supply.

The secondary coil of the transformer has 418 turns of wire.

c Calculate the number of turns on the primary coil of the transformer.

d If the transformer was incorrectly wired and the 418-turn coil was connected to the mains supply, what voltage would there be at the output?

Electronics – input/output devices

1 In which of the following circuits will the LED light up?

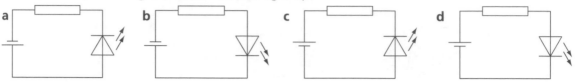

2 Calculate the missing quantities for the circuits below. In each circuit, the LED is rated at 1·8 V.

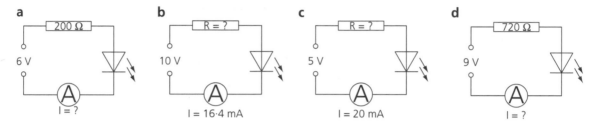

3 For the amplifier circuits shown below, calculate the missing quantities.

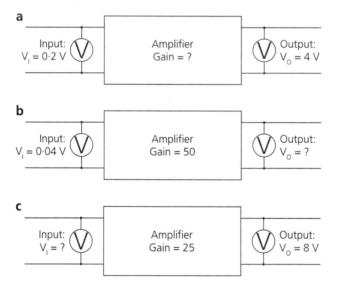

a

Input: $V_I = 0.2$ V   Amplifier Gain = ?   Output: $V_O = 4$ V

b

Input: $V_I = 0.04$ V   Amplifier Gain = 50   Output: $V_O = ?$

c

Input: $V_I = ?$ Amplifier Gain = 25   Output: $V_O = 8$ V

4 Complete the following table for amplifiers using the appropriate fomula:

Input voltage	Output voltage	Gain
0·8 V	16 V	
	9 V	30
2·2 V		15
20 mV	6 V	
	3 V	4
3 mV		200

See also Leckie & Leckie's Intermediate 2 Physics Course Notes

A rear bicycle light uses the circuit shown in the diagram below.

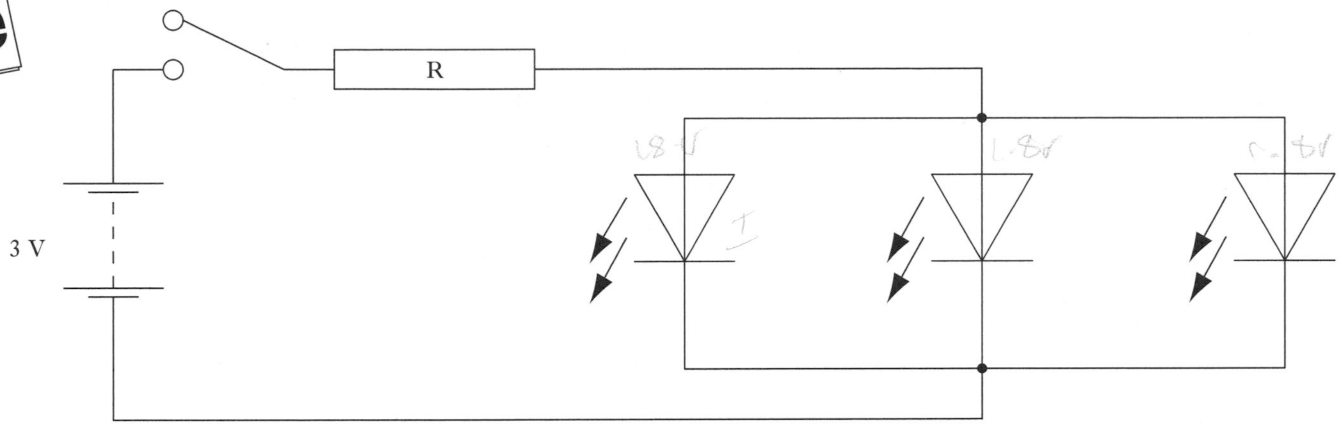

Each of the LEDs is designed to operate with a potential difference of 1·8 V and a current of 20 mA.

a   Calculate the current flowing through the resistor when the LEDs are operating correctly.

b   Calculate the potential difference across the resistor when the LEDs are operating correctly.

c   Calculate the value of resistor that is required for this circuit to operate correctly.

d   State one possible reason why the LEDs have been put in parallel rather than in series.

See Answers on page 7 of answer booklet

Transistor circuits

1   Calculate the voltage across the thermistor in each of the following circuits.

2   Calculate the voltage across the LDR (light dependent resistor) in each of the following circuits.

3   This circuit allows the voltage across the base and emitter of the transistor to be varied.
    Complete the table to show whether the transistor and LED are ON or OFF.

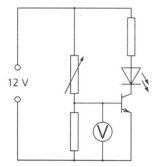

V (V)	Transistor	LED
1·4	ON	
0·4		
0·9		
0·3		

4   This circuit allows the voltage across the gate and source to be varied. Complete the table to show
    if the MOSFET and lamp are ON or OFF.

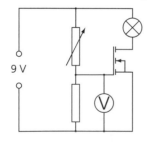

V (V)	MOSFET	Lamp
0·8		
2·4	ON	
1·3		
3·2		

See also Leckie & Leckie's Intermediate 2 Physics Course Notes

Maggie, a school pupil, designs her own home burglar alarm system containing a time delay circuit which allows 30 seconds for a secret code to be punched into a keypad after the front door has been opened. If the code has not been entered correctly within 30 seconds, the output from the time delay circuit changes from 0·5 V to 11 V which triggers the alarm. Part of her circuit is shown below.

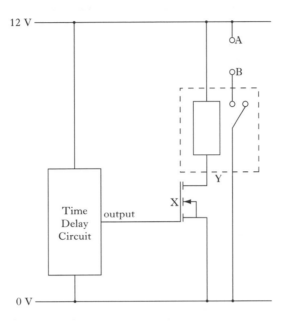

Component X in the circuit is represented by the symbol shown below

a Identify components X and Y.

b What do the letters g, d and s represent?

c Maggie wants to have both a 12 V siren and a 12 V light bulb to warn when the alarm is triggered. Draw a diagram to show how the two components should be connected between points A and B.

d Explain how the alarm is made to operate by the actions of components X and Y.

4 Waves and Optics

Basic wave ideas

1 Calculate the missing quantities in the table using the appropriate formula.

Speed	Frequency	Wavelength	Period
	5 Hz	2 m	
20 m/s	10 Hz		
		3 m	0·1 s
340 m/s		17 m	
320 m/s	8 kHz		
		6 m	0·004 s

The diagram shows part of a new wave pool in a Spanish water park.

A wave generator makes waves at a rate of 30 waves per minute.

a State whether water waves are transverse or longitudinal

b Use the diagram to estimate
   i the amplitude of the waves
   ii the wavelength of the waves.

c Calculate the frequency of the waves in hertz.

d Calculate the period of the waves in seconds.

e Calculate the speed of the waves.

f How long would a wave crest take to travel from A to B?

See also Leckie & Leckie's Intermediate 2 Physics Course Notes

Reflection

1 Which of the following reflected rays is at the correct angle?

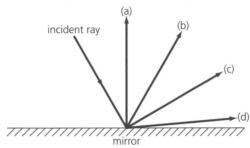

The diagram below shows how a ray of visible light reflects when it hits a mirror.

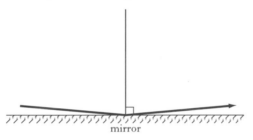

a On the diagram label the incident ray, reflected ray, normal, angle of incidence and angle of reflection

b Two communication systems send data from a transmitter to a receiver. One system does this with a beam of microwaves travelling in air and the other uses light travelling in an optical fibre. Complete the diagram below to show what happens to the beam of microwaves that are travelling towards the curved reflector.

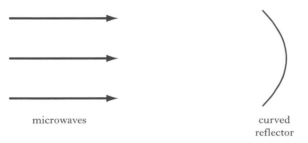

microwaves   curved
                                            reflector

c Where should the detector of microwaves be placed to detect the strongest signal?

d The diagram below shows a short length of optical fibre from the second communication system. Complete the diagram to show how the ray of light travels through this optical fibre.

Refraction

1 Which of the following refracted rays follows the correct path?

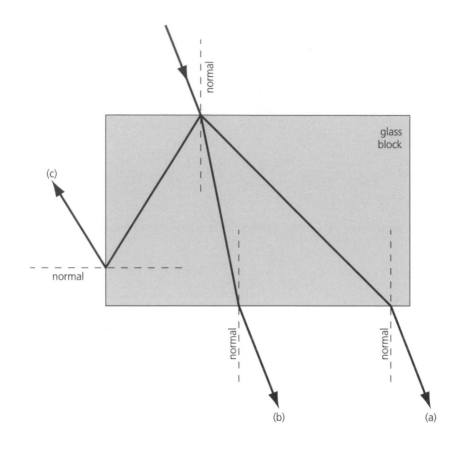

2 Copy the diagrams and complete them showing the path of the light rays through the glass blocks.

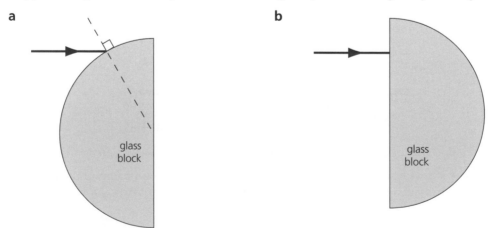

See also Leckie & Leckie's Intermediate 2 Physics Course Notes

A fishpond has a submerged light at the bottom which can be switched on after dark. The critical angle for water is 49°. The diagram shown below shows 3 particular rays of light travelling from the light towards the surface of the water.

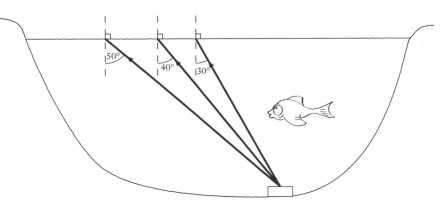

a Complete the diagram to show the path that each ray of light takes when it reaches the surface of the water.

The diagram shown below illustrates how the objective lens and eyepiece lens in a pair of binoculars can be offset from each other using two triangular glass prisms. The path of one ray of light is shown.

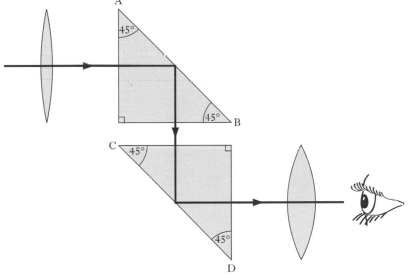

b The prisms are clear glass and do not have silvered surfaces. What name is given to reflection of this type within a glass block?

c The angle of incidence on faces AB and CD of the prisms is 45°. From this information, what can be said about the critical angle for glass?

Lenses – the eye

1 Complete the ray diagrams for these lenses. The focal length for each lens is indicated.

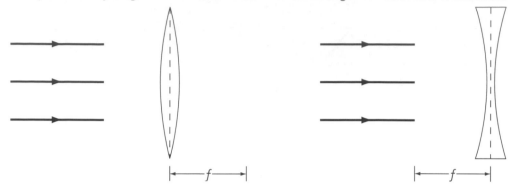

2 The following table has data for concave and convex lenses. Complete the table using the appropriate formula.

Lens type	Focal length	Power
Convex	0·2 m	
Convex		10 D
		-4 D
Concave	40 cm	
		2 D
		-5 D

Copy and complete each of the three ray diagrams shown below to show where the image would be formed, then use three terms from the following list to describe each of the images.

Real, Virtual, Inverted, Upright, Diminished, Magnified, Same size

a

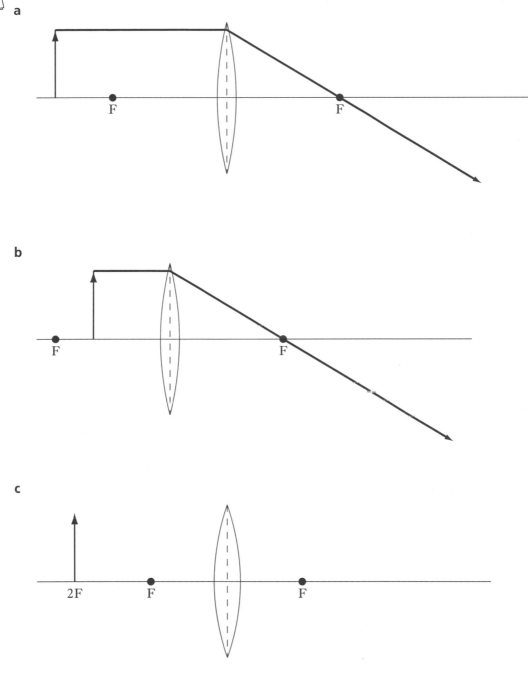

b

c

5 Radioactivity

Properties of alpha, beta, gamma

1 The diagram shows the penetration of three different kinds of radiation (alpha, beta and gamma) through a series of absorbers.

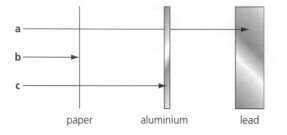

Identify each of the radiations labelled a, b, and c.

2 This diagram shows the paths of five different radiations labelled P, Q, R, S and T passing between two electrically charged plates.

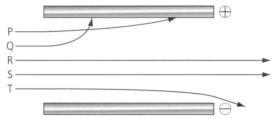

The radiations are alpha particles, slow moving beta particles, neutrons, fast beta particles and gamma rays.

a Identify radiation T.

b Explain why it is not possible to distinguish between neutron and gamma ray paths.

c Explain which path is caused by the fast beta particles.

3 A nucleus of uranium which contains 92 protons and 146 neutrons emits a particle X. The resulting daughter nucleus contains 90 protons and 144 neutrons. The equation from the decay can be written as

$$^{238}_{92}\text{U} \longrightarrow {}^{234}_{90}\text{Th} + \text{X}$$

Identify the particle X.

4 A radioactive carbon-14 nucleus decays according to the following equation.

$$^{14}_{6}\text{C} \longrightarrow {}^{14}_{7}\text{C} + \text{Y}$$

Identify particle Y.

5 Sulphur-31 decays into phosphorus and emits a unusual particle, labelled Z below, you have probably not heard of.

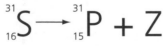

$$^{31}_{16}\text{S} \longrightarrow {}^{31}_{15}\text{P} + \text{Z}$$

Try to describe this 'new' particle!

See also Leckie & Leckie's Intermediate 2 Physics Course Notes

A scientist designs a system to monitor the thickness of long sheets of a plastic flooring material as they pass over rollers on a production line. A radioactive source is placed above the sheets and radiation which passes through the sheets is detected underneath them, using a detector and a counter. The system should work effectively as long as the activity of the source remains above 20 MBq.

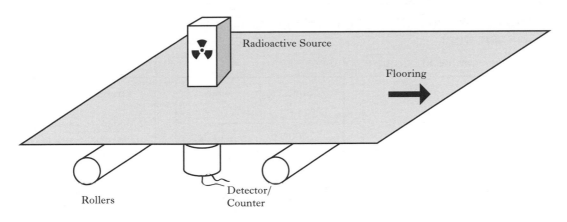

a Explain what should happen to the reading on the counter if a faulty batch of flooring is too thin.

Details of three radioactive sources available to the scientist are shown on the table below.

Source	Type	Activity (MBq)	Half-life
A	Alpha	100	460 years
B	Beta	80	28 years
C	Beta	320	62 days

b Explain why source A is totally unsuitable.

c Explain which source is the best to use.

Dosimetry

1 A Geiger-Müller tube is connected to the electronic counters and timers shown below to measure the activities of various radioactive sources.

   a Describe briefly how a Geiger-Müller tube works. You should include the word 'ionisation' in your description

   b Find the activities in becquerels of these sources.

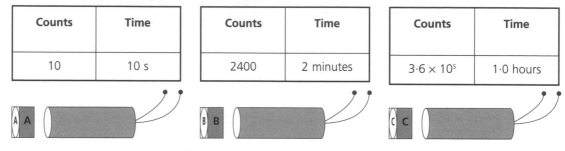

Counts	Time
10	10 s

Counts	Time
2400	2 minutes

Counts	Time
$3 \cdot 6 \times 10^5$	$1 \cdot 0$ hours

2 A source has an activity of 2 kBq. How many decays will occur in

   a 5 seconds

   b 100 seconds

   c 20 minutes

   d 1 hour?

3 Find the absorbed doses in grays, when

   a a mass of 10 kg absorbs 2 J of energy

   b a mass of 500 g absorbs $0 \cdot 25$ J of energy

   c a mass of 10 g absorbs 2 mJ of energy

   d a mass of 50 kg absorbs $0 \cdot 2$ kJ of energy.

4 The diagrams below each show the absorption of 30 mJ of energy from different radioactive sources.

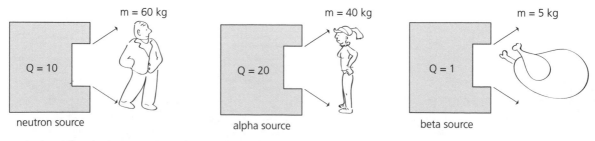

m = 60 kg

Q = 10

neutron source

m = 40 kg

Q = 20

alpha source

m = 5 kg

Q = 1

beta source

Calculate the dose equivalent in sieverts in each case.

See also Leckie & Leckie's Intermediate 2 Physics Course Notes

A research scientist is investigating radiation doses received by a number of different people. Some of his data appear in the table below.

a   What is the unit of absorbed dose?

b   What is the unit of dose equivalent?

c   Complete the table showing absorbed dose and dose equivalent for each of the three subjects.

	Mass exposed to radiation	Energy absorbed	Absorbed dose	Radiation type	Quality factor	Dose equivalent
Medical physics technician	Whole body 70kg	0·245 J		Beta particles	1	
Nuclear power station worker	Whole body 60kg	0·015 J		Neutrons	10	
Hospital patient receiving radiotherapy	Tumour 5g	0·075 J		Gamma rays	1	

d   People can be protected from overexposure to radiation by the use of heavy metal or concrete shielding. Name another way in which radiation dosage can be minimised.

5 Radioactivity

Half-life

1 Describe the meaning of the term 'half-life'.

2 Phosphorus-33 has a half-life of 25 days. A sample with an activity of 100 Bq is left for 75 days. Find its activity after that time.

3 Molybdenum-101 has a half-life of 15 minutes. What mass of molybdenum-101 would remain if 1200 g was left for 1 hour?

4 A scientist finds that the activity of a piece of radioactive material is 1 kBq. He comes back 3 days later and finds that the activity has fallen to 125 Bq. Find the half-life of the material in days.

5 The half-life of an isotope is being measured while it is in highly radioactive surroundings. The background reading of the surroundings remains at 50 Bq. Use the table of results below to find the half-life of the isotope.

Total Activity (Bq)	250	190	150	120	100	85
Times (minutes)	0	5	10	15	20	25

6 The radioactive isotope potassium-40 has a half-life of 1.25×10^9 years. When the Earth became a solid planet, a rock formed containing a small trace (2 mg) of potassium-40. Today, the rock remains intact but there is only 0·5 mg of the potassium-40 remaining in it. Use this information to estimate the age of the Earth.

a A scientist has been recording the activity a new radioactive substance X. Her data are shown in the table below. She concludes that the half-life is 60 minutes. Draw a graph of activity against time and then use it to check her conclusion. Is her conclusion correct?

Time (minutes)	Activity of substance X (Bq)
0	368
30	248
60	182
90	115
120	90
150	58
180	42
210	26
240	20
270	17

b A larger sample of substance X has an initial activity of 8 kBq. Find the activity of this sample after six hours.

c The scientist is given a piece of similar-looking radioactive material by a colleague at 12 noon. Its activity is 5 kBq at that time. The colleague claims that when he measured its activity at 9 a.m. that morning the activity was 60 kBq. Explain whether this sample could also be substance X.

5 Radioactivity

Nuclear reactors

1   The diagram shows the fission of a uranium 235 nucleus.

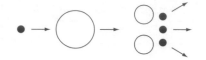

Copy the diagram and add the following labels:
Slow neutron   fission fragments   fast neutrons   uranium-235 nucleus.

2   The two diagrams below illustrate a controlled fission chain reaction in a power station and an uncontrolled chain reaction in a bomb.

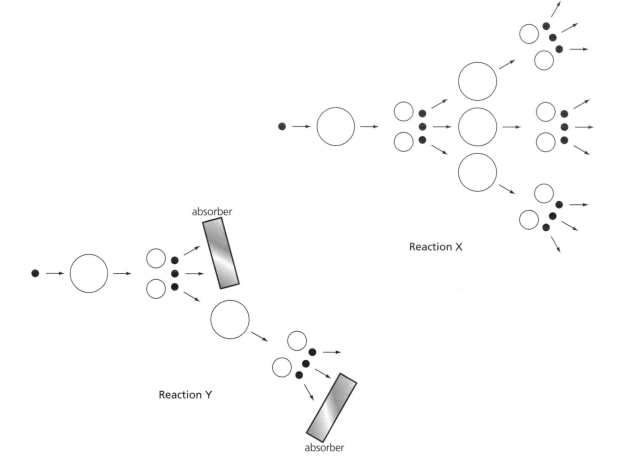

Reaction X

absorber

Reaction Y

absorber

Explain which reaction is the bomb reaction.

3   In reaction X, the 'first generation' fission produces three neutrons and the 'second generation' produces nine neutrons. Predict how many neutrons will have been produced by the 'fifth generation'.

4   Imagine you are 'antinuclear' campaigner. Write a short paragraph on the 'evils of nuclear power'. Mention at least two disadvantages of nuclear power.

5   You are now a 'pro nuclear' campaigner. Write a brief note on the benefits of nuclear power. Mention at least two advantages.

See also Leckie & Leckie's Intermediate 2 Physics Course Notes

The diagram shows a simplified version of the core of one of the two reactors in Torness Nuclear Power Station.

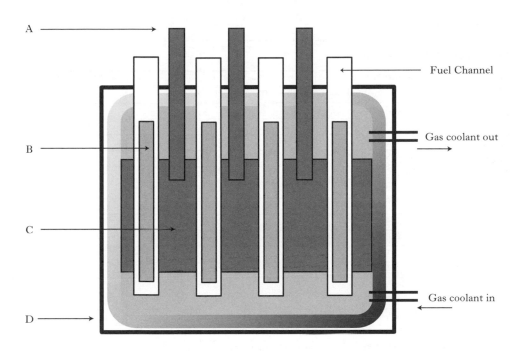

A

B

C

D

Fuel Channel

Gas coolant out

Gas coolant in

The station uses the heat produced by nuclear fission to generate electricity.

a Write a simple description of nuclear fission using a diagram if you wish.

b Some parts of the reactor core are already labelled. Copy and complete the diagram by identifying parts A, B, C and D.

c Describe briefly the functions of parts A, B, C and D.

d Each of the two reactors in Torness produces 1500 MW of heat. The station sends a total of 1200 MW of electrical power to the grid. Calculate the percentage efficiency of the station.

e Another power station, Longannet, produces 2400 MW of electricity by burning coal. State one advantage and one disadvantage of this method as compared to nuclear generation.

6 Useful Physics Data

Prefixes

Symbol	Name	Represents
μ	micro	10^{-6}
m	milli	10^{-3}

Symbol	Name	Represents
k	kilo	10^{3}
M	Mega	10^{6}
G	Giga	10^{9}

Abbreviations and Units

Symbol	Meaning	Units	Abbreviated Units
v	average (or final) speed	metres per second	m/s
s	distance	metres	m
t	time	seconds	s
a	acceleration	metres per second per second	m/s²
u	initial speed	metres per second	m/s
w	weight	newtons	N
m	mass	kilograms	kg
g	gravitational field strength	newtons per kilogram	N/kg
F	force	newtons	N
	momentum	kilogram metres per second	kg m/s
E_w	work done	joules	J
E	energy	joules	J
P	power	watts	W
E_p	gravitational potential energy	joules	J
E_k	kinetic energy	joules	J
E_h	heat transferred	joules	J
c	specific heat capacity	joules per kilogram degree celsius	J/kg°C
ΔT	change in temperature	degrees celsius	°C
l	specific latent heat	joules per kilogram	J/kg
Q	charge	coulombs	C
I	current	amperes (amps)	A
V	voltage (potential difference)	volts	V
R	resistance	ohms	Ω
λ	wavelength	metres	m
f	frequency	hertz	Hz
P	lens power	dioptres	D
f	lens focal length	metres	m
N	number of decays	no units	
A	activity	becquerels	Bq
H	dose equivalent	sieverts	Sv
D	absorbed dose	grays or joules per kilogram	Gy or J/kg
Q	quality factor	no units	

Equations for Mechanics and Heat

$v = \frac{s}{t}$

$F = ma$

$E = Pt$

% Efficiency $= \frac{\text{useful energy out}}{\text{total energy in}} \times 100$

$v = u + at$

momentum $= mv$

$E_p = mgh$

$E_h = cm\Delta T$

$w = mg$

$E_w = Fs$

$E_k = \frac{1}{2}mv^2$

$E_h = ml$

Equations for Electricity and Electronics

$Q = It$

$P = I^2R$

$V = IR$

$P = \frac{V^2}{R}$

$P = IV$

$E = Pt$

Series resistors: $R_t = R_1 + R_2 + R_3$

Parallel resistors: $\frac{1}{R_t} = \frac{1}{R_1} + \frac{1}{R_2} + \frac{1}{R_3}$

Equations for Waves and Optics

$V = f\lambda$

$P = \frac{1}{f}$

Equations for Radioactivity

$A = \frac{N}{t}$

$H = DQ$

$D = \frac{E}{m}$

Answers to Leckie & Leckie's
Questions in Intermediate 2 Physics

Scalars and vectors

1 **a** 4 m/s
    b 1800 m
    c 20 s

2

Runner	Race Distance	Time	Average Speed
B. Bloggs	100 m	12·5 s	8 m/s
H. Gebresellasie	5000 m	800 s (13 min 20 s)	6·25 m/s
S. Motion	1500 m	8 min 20 s	3 m/s

3 **a** 35,000 m
    b 20,000 m
    c 48,000 m

4 **a** 2 s
    b 100,000 s (27 hours 46 minutes 40 s)
    c 500 s (8 minutes 20 s)

5 **a** **i** 22 m   **b** **i** 10 m (090°)
      ii 7 m   **ii** 5 m (053°)
      iii 34 m   **iii** 26 m (113°)

6 **i** 2·2 m/s, 1 m/s (090°)
    ii 0·7 m/s, 0·5 m/s (053°)
    iii 3·4 m/s, 2·6 m/s (113°)

7 **a** 0·5 m/s
    b 1 m/s

E **a** $v = \frac{s}{t} = \frac{5 \cdot 5}{11} = 0 \cdot 5$ m/s

    b $s = vt = 0 \cdot 5 \times 6 = 3$ m

    c 2 m directly to the right of the starting point

    d $v = \frac{s}{t} = \frac{2}{11} = 0 \cdot 18$ m/s directly to the right of the starting point

    e $v = \frac{s}{t} = \frac{2}{6} = 0 \cdot 33$ m/s directly to the right of the starting point

Projectiles

1 **a** 18 m
    b 36 m

2 **a** 15 m/s
    b 30 m/s

3 **a** 15 m/s
    b 30 m/s
    c 36 m
    d

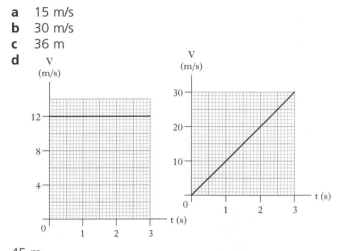

45 m
    e 32·3 m/s at 68 degrees below horizontal

E **a**

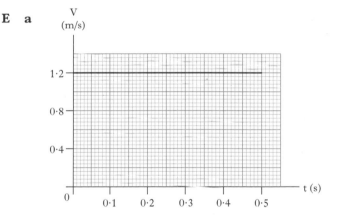

    b $s = vt = 1 \cdot 2 \times 0 \cdot 5 = 0 \cdot 6$ m

    c

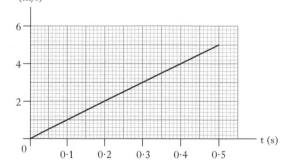

    d distance = area below graph
         = $0 \cdot 5 \times 5 \times 0 \cdot 5 = 1 \cdot 25$ m

    e $s = vt = 2 \times 0 \cdot 5 = 1$ m
        As distance XY was 0·6 m, the bomb now lands 0·4 m beyond Y.

Graphs of motion

1 **a** 10 m/s²
  b 8 m/s²
  c −2 m/s²

2 **a** 48 m
  b 125 m
  c 125 m

3 **a**

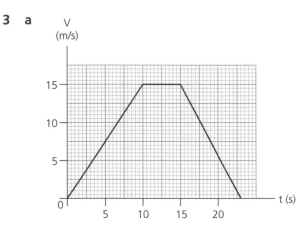

  b 1·5 m/s²
  c 5 m/s²
  d 247·5 m
  e 10·3 m/s

4 **a** between 15 s and 16 s
  b 181·5 m
  c 8·6 m/s

E **a** $a = \frac{v-u}{t}$
      $= \frac{25-0}{5}$
      $= 5 \text{ m/s}^2$

  b point B is reached after 5s
      distance = area below graph
          $= 0{\cdot}5 \times 5 \times 25$
          $= 62{\cdot}5$ m

  c 20 m/s

  d distance = area below graph
          $= (2 \times 20) + (0{\cdot}5 \times 2 \times 5)$
          $= 45$m

  e total distance travelled $= 62{\cdot}5 + 45$
              $= 107{\cdot}5$ m

      $v = \frac{s}{t}$
        $= \frac{107{\cdot}5}{7}$
        $= 15{\cdot}4$ m/s

Forces

1 **a** 600 N   **b** 37·5 kg   **c** 3 N/kg

2

Weight (N)	Mass (kg)	Gravitational Field Strength (N/kg)
100	5·0	20
140	20	7·0
200	0·8	250
400	50·0	8·0
32	2	16·0
90	75·0	1·2

3 0·05 kg (50 grammes)

4 3·9 N

5 **a** 50 N   **b** 8 N

6 **a** 2 m/s²
  b 1250 kg
  c 4000 N

7

Force (N)	Mass (kg)	Acceleration (m/s²)
100	50	2
45	3	15
100	25	4
3000	300	10
270	900	0·3
4	20	0·2

8 0·63 m/s²

9 12,500 kg

E **a** w $= mg$
        $= 2,000,000 \times 10$
        $= 20,000,000$ N

  b Total thrust
      $= (3 \times 1,700,000) + (2 \times 12,000,000)$
      $= 29,100,000$ N

  c Resultant Force
      $= 29,100,000 - 20,000,000$
      $= 9,100,000$ N

  d $a = \frac{F}{m}$
      $= \frac{9,100,000}{2,000,000}$
      $= 4{\cdot}55$ m/s²

  e $a = \frac{F}{m}$
      $= \frac{(3 \times 3000)}{75,000}$
      $= 0{\cdot}12$ m/s²

  f Friction due to the air

Momentum and collisions

1 **a** 275 kg m/s
   b 1·5 kg
   c 7 m/s

2

Momentum (kg m/s)	Mass (kg)	Velocity (m/s)
270	90	3
500	20	25
240	60	4
16,500	1100	15
3000	2000	1·5
120	12	10

3 6 kg m/s

E **a** $s = vt = 3 \times 10^8 \times 1 \times 10^{-6} = 300$ m
    300 m is the total distance travelled by the
    microwaves, therefore the car is half this
    distance away, i.e. 150 m

   b **i** momentum = mv
            = $(3000 + 1000) \times 25$
            = 100,000 kg m/s

     ii momentum is conserved during the
        collision so momentum before was also
        100,000 kg m/s

    iii total momentum before
         = total momentum after
         $(3000 \times 30) + 1000v = 100,000$
         $v = \frac{100,000 - 90,000}{1000}$
         $v = 10$ m/s

Energy

1 **a** 1600 J
   b 50 m
   c 400 N

2

Work Done (J)	Force (N)	Distance (m)
150	50	3
300	4	75
4500	30	150
180,000	150	1200
3000	200	15
120,000	600	200

3 **a** 27,000 J
   b 4 m
   c 5 kg

4

Potential Energy (J)	Mass (kg)	Height (m)
480	15	3·2
150	1·25	12
250	10	2·5
21	0.75	2·8
730,400	80	913
234,000	1300	18

5 90 m

6 0·8 m

7 **a** 812·5 J
   b 8 kg
   c 3 m/s

8

Kinetic Energy (J)	Mass (kg)	Speed (m/s)
3442·5	85	9
196	200	1·4
32	1	8
80,275	950	13
93750	3	250
750	3000	0·71

9 6·4 J

10 30 m/s

E **a** $E_k = \frac{1}{2}mv^2 = 0.5 \times 60 \times 1^2 = 30$ J

   b $E_p = mgh = 60 \times 10 \times 3.15 = 1890$ J

   c Total energy
     = Potential energy + Kinetic energy
     = 30 + 1890
     = 1920 J

   d All energy is now kinetic.
     $E_k = \frac{1}{2}mv^2$
     $v = \sqrt{\frac{2E_k}{m}}$
       $= \sqrt{\frac{2 \times 1920}{60}}$
       $= 8$ m/s

   e $F = \frac{E_w}{s}$
      $= \frac{1920}{20}$
      $= 96$ N

Specific/latent heat

1 **a** 33,600 J
   b 58,800 J
   c 75,600 J

2

Initial temp (°C)	Final temp (°C)	Mass (kg)	Energy supplied (J)	Specific heat capacity (J/kg°C)
10	20	5	50,000	1000
100	200	8	3,200,000	4000
16	17	4	6,000	1500
50	52	10	84,000	4200
150	250	3	39,000	130

3 1454 J/kg°C

4 **a** 4×10^6 J
   b $1 \cdot 2 \times 10^6$ J
   c $2 \cdot 4 \times 10^5$ J

5 Ice: 334,000 J/kg
   Lead: 24,741 J/kg

6 1,550,000 J/kg

E **a** Total mass $= 6 \times 0 \cdot 3 = 1 \cdot 8$ kg
     $E = cm\Delta T$
       $= 4000 \times 1 \cdot 8 \times (20 - 5)$
       $= 108,000$ J

   b $t = \frac{E}{P}$
       $= \frac{108,000}{100}$
       $= 1080$ s

   c Energy to cool from 20°C to 0°C
     $E = cm\Delta T$
       $= 4000 \times 0 \cdot 3 \times 20$
       $= 24,000$ J

     Energy to freeze solid
     $E_h = ml$
       $= 0 \cdot 3 \times 300,000$
       $= 90,000$ J

     Total energy $= 24,000 + 90,000$
                  $= 114,000$ J

     $t = \frac{E}{P}$
       $= \frac{114,000}{57}$
       $= 2000$ s

Mechanical energy/heat

1 Metal X: 500 J/kg°C
   Metal Y: 800 J/kg°C

2 Lump A: 12°C
   Lump B: 2°C

3 20°C

4 260°C

E **a** $E_k = \frac{1}{2}mv^2$
       $= 0 \cdot 5 \times 850 \times 30^2$
       $= 382,500$ J

   b New $E_k = 0 \cdot 5 \times 850 \times 122$
             $= 61,200$ J
     E_k lost $= 382,500 - 61,200$
             $= 321,300$ J

   c Kinetic energy to heat energy

   d **i** $\Delta T = \frac{E}{cm}$
           $= \frac{321,300}{700 \times 7 \cdot 6}$
           $= 60 \cdot 4$°C

      ii Final temperature
        = Initial temperature + temperature rise
        $= 25 + 60 \cdot 4$
        $= 85 \cdot 4$°C

   e Some heat energy is transferred to the surroundings during the period of braking so not all of the heat energy stays in the brake discs.

Electrostatics

1 **a** 30 C
   b 90 s
   c 0·5 A

2

Current	Charge	Time
10 A	45 C	4·5 s
3 A	180 C	60 s
2·3 C	46 C	20 s
0·25 A	1200 C	4800s
5 A	6 C	1·2 s
1·5 A	135 C	90 s

3 1·2 C

4 10 A

E **a** electrons

   b Atoms have a positive nucleus surrounded by negative electrons. Overall the charge of an atom is neutral because the charge on the electrons cancel out the charge on the nucleus. Some electrons have been transferred from the atoms of the seat to Ralph making him negatively charged. The atoms of the seat no longer have sufficient electrons to balance the positive charge of the nuclei so the seat has become positively charged

   c $I = \frac{Q}{t}$
$$= \frac{6 \times 10^{-6}}{1 \times 10^{-3}}$$
$$= 6 \times 10^{-3} \text{ A}$$

   d 1 hour is $60 \times 60 = 3600$ s
      $Q = It$
$$= 36 \times 3600$$
$$= 129{,}600 \text{ C}$$

   e light energy to electrical energy

   f $t = \frac{Q}{I}$
$$= \frac{129{,}600}{0·05}$$
$$= 2{,}592{,}000 \text{ s}$$
      8 hours $= 8 \times 60 \times 60$
$$= 28{,}800 \text{ s}$$
      number of days $= \frac{2{,}592{,}000}{28{,}800}$
$$= 90 \text{ days}$$

Circuits – series/parallel

1 $A_2 = 2$ A, $V_2 = 6$ V   **2** $V_2 = 6$ V

3 $V_S = 12$ V

4 $A_3 = 3$ A, $V_1 = 9$ V

5 $A_2 = 1·8$ A

6 $V_S = 6$ V, $A_2 = 0·5$ A

7 $3\,\Omega$ **8** 3 V   **9** $0·6$ A

10 **a** $50\,\Omega$ **b** $2·5$ kΩ
    c $5·5\,\Omega$ **d** $167\,\Omega$
    e $9\,\Omega$

11 $354\,\Omega$

12 50 A

E **a** 60 cm = 0·6 m
    Resistance of X = $0·6 \times 20 = 12\,\Omega$

   b 100 cm = 1 m
    Resistance of Y = $1 \times 20 = 20\,\Omega$

   c $R_t = R_1 + R_2 + R_3 = 12 + 12 + 20 = 44\,\Omega$

   d $\frac{1}{R_t} = \frac{1}{R_1} + \frac{1}{R_2} = \frac{1}{12} + \frac{1}{12}$ $R_t = 6\,\Omega$

   e $\frac{1}{R_t} = \frac{1}{R_1} + \frac{1}{R_2} = \frac{1}{12} + \frac{1}{20}$ $R_t = 7·5\,\Omega$

   f R_t for X and Y in series is $12 + 20 = 32\,\Omega$
    For parallel combination:
    $\frac{1}{R_t} = \frac{1}{R_1} + \frac{1}{R_2} = \frac{1}{32} + \frac{1}{32}$ $R_t = 16\,\Omega$

   g R_t for X and Y in parallel = $7·5\,\Omega$
    Voltmeter reads 15V which is the supply voltage.
    $I = \frac{V}{R} = \frac{15}{7·5} = 2A$ The ammeter reads 2 A.

   h For resistance of parallel section
    $\frac{1}{R_t} = \frac{1}{R_1} + \frac{1}{R_2} = \frac{1}{20} + \frac{1}{20}$ $R_t = 10\,\Omega$

    For resistance of whole circuit
    $R_t = R_1 + R_2 = 10 + 20 = 30\,\Omega$

    For current flowing from supply
    $I = \frac{V}{R} = \frac{15}{30} = 0·5$ A

    For voltmeter reading
    $V = IR = 0·5 \times 20 = 10$ V
    The voltmeter reads 10 V.

    For ammeter reading
    The 0·5 A flowing from the supply splits equally through the parallel section as both resistors have the same resistance.
    The ammeter reads half of 0·5 A = 0·25 A

Circuits – power

1 **a** 18 W
  b 1200 W
  c 0·25 W

2 **a** 2 A
  b 0·2 A
  c 0·42 A

3 **a** 100 V
  b 10 V
  c 12 V

4 **a** 10 Ω
  b 100 Ω
  c 100 Ω

E **a** Off is position A

  b Normal power is position C

  c $I = \frac{P}{V}$
    $= \frac{12}{6}$
    $= 2$ A

  d $R = \frac{V}{I}$
    $= \frac{6}{2}$
    $= 3\ \Omega$

  e $P = I^2R$
    $= 1 \cdot 5^2 \times 3$
    $= 6 \cdot 75$ W

  f Advantage: The battery lasts longer.
    Disadvantage: The lamp is not as bright.

Oscilloscope – AC/DC

1 **a** 1 V
  b 5 V/division

E **a** The yellow terminals as the voltage cycles both positive and negative.

  b The electrons move forwards and backwards and forwards etc.

  c The effective AC voltage of the yellow terminals is 6 V which is the same as the DC voltage from the red and black terminals.

  d The manufacturers were labelling the yellow terminals with their effective voltage.

Electromagnetism

1 **a** coil R
  b coil P

2 **a** 0·9 V
  b 1·8 V
  c 3·6 V
  d 0 V
  e 5·4 V

E **a** The movement of the stylus in the recford groove causes the magnet to move near the coil. This causes an induced voltage.

  b a.c.

  c This should cause the magnet to move faster, generating a larger voltage.

  d stronger magnet
    more turns of wire on the coil

Transformers

1 **a** 20 V
  b 6·0 V
  c 2 V

2 **a** 0·25 A
  b 0·2 A

3 **a** $I_S = 2$ A, $I_P = 24$ A
  b $I_S = 1 \cdot 2$ A, $I_P = 0 \cdot 012$ A
  c $I_S = 6$ A, $I_P = 180$ A

E **a** $I = \frac{P}{V} = \frac{1000}{110} = 9 \cdot 1$ A

  b power in primary = power in secondary
    $\Rightarrow 1000 = V_P I_P \Rightarrow I_P = \frac{1000}{230} = 4 \cdot 3$ A

  c $\frac{V_S}{V_P} = \frac{N_S}{N_P} \Rightarrow N_P = \frac{N_S V_P}{V_s} = \frac{418 \times 230}{110} = 874$ turns

  d $\frac{V_S}{V_P} = \frac{N_S}{N_P} \Rightarrow V_P = \frac{N_S V_P}{N_P} = \frac{874 \times 230}{418} = 481$ V

Electronics – input/output devices

1 The LED will light up in circuits **a** and **d**.

2 **a** $I = 0.021$ A
   b $R = 500\ \Omega$
   c $R = 160\ \Omega$
   d $I = 0.01$ A

3 **a** Gain = 20
   b $V_O = 2$ V
   c $V_I = 0.32$ V

4

Input voltage	Output voltage	Gain
0·8 V	16 V	20
0·3 V	9 V	30
2·2 V	33 V	15
20 mV	6 V	300
0·75 V	3 V	4
3 mV	0·6 V	200

E **a** $I = 3 \times 20$ mA $= 60$ mA

   b $V = 3 - 1.8 = 1.2$ V

   c $R = \dfrac{V}{I} = \dfrac{1.2}{0.06} = 20\ \Omega$

   d If one fails, the others will continue to operate.

Transistor circuits

1 **a** 7 V
   b 1 V

2 **a** 1·7 V
   b 1 V

3

V (V)	Transistor	LED
1·4	ON	ON
0·4	OFF	OFF
0·9	ON	ON
0·3	OFF	OFF

4

V (V)	MOSFET	Lamp
0·8	OFF	OFF
2·4	ON	ON
1·3	OFF	OFF
3·2	ON	ON

E **a** X = MOSFET
   Y = relay switch

   b g = gate
   d = drain
   s = source

   c
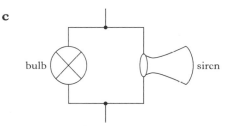

   d When the time delay circuit changes to 11 V, the MOSFET 'switches on'. Current flows through the coil of the relay R, pulling the switch S on and allowing a current to flow through the bulb and siren.

Basic wave ideas

1

Speed	Frequency	Wavelength	Period
10 m/s	5 Hz	2 m	0·2 s
20 m/s	10 Hz	2 m	0·1 s
30 m/s	10 Hz	3 m	0·1 s
340 m/s	20 Hz	17 m	0·05 s
320 m/s	8 kHz	0·04 m	$1·25 \times 10^{-4}$ s
1500 m/s	250 Hz	6 m	0·004 s

E a Transverse

b i amplitude $= \frac{1}{2} \times 1· 5 = 0·75$ m

ii wavelength $= \frac{12}{3} = 4$ m

c $f = \frac{30}{60} = 0·5$ Hz

d $T = \frac{1}{f} = \frac{1}{0·5} = 2$ s

e $v = f\lambda = 0·5 \times 4 = 2$ m/s

f $t = \frac{d}{v} = \frac{d}{v} = 6$ s

Reflection

1 Ray b

E a

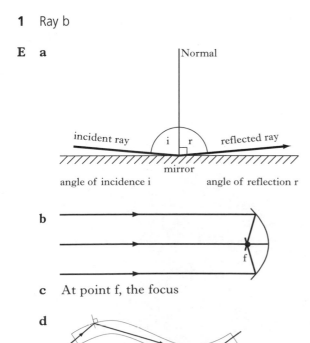

c At point f, the focus

d

Refraction

1 path b

2 a

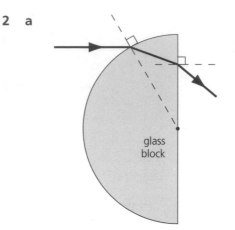

b

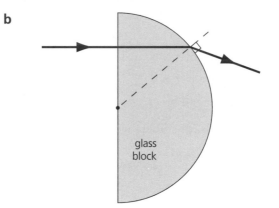

E a

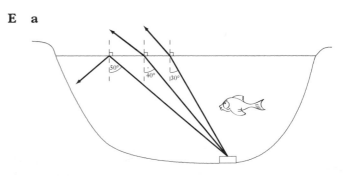

b Total internal reflection

c It must be less than 45°.

Lenses – the eye

1

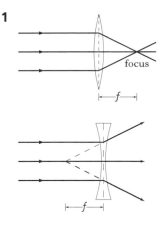

2

Lens type	Focal length	Power
Convex	0·2 m	5 D
Convex	0·1 m	10 D
Concave	0·25 m	−4 D
Concave	40 cm	−2·5 D
Convex	0·5 m	2 D
Concave	0·2 m	−5 D

E a

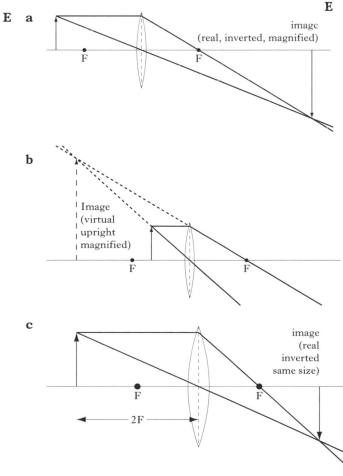

image (real, inverted, magnified)

b

Image (virtual upright magnified)

c

image (real inverted same size)

Properties of alpha, beta, gamma

1 **a** gamma

   b alpha

   c beta

2 **a** T = alpha

   b Both have no electrical charge. They are not deflected by charged plates.

   c Path P. Paths P and Q are both beta particle paths. The negatively charged beta particles are attracted to the positively charged plate but the faster beta particles will go further before they hit the plate.

3 Particle X must be $^4_2$He, an alpha particle. You could also describe it as a helium nucleus.

4 Particle Y must be $^0_{-1}$e, a beta particle.

5 Particle Z must be $^0_1$e. It is the same size as an electron but it has a positive charge. It is called a positron. It is the 'antiparticle' of the beta particle.

E a If the flooring is too thin, more radiation would pass through it, causing an increased count reading on the counter.

   b Source A is an alpha source. Alpha particles can be stopped by paper or even a few cm of air. It is unlikely that any alphas would be able to pass through the flooring to register on the counter.

   c Source B is the best source. Beta particles will pass through flooring sufficient numbers for the system to work. The long half-life will mean that the source will hardly ever need to be replaced.

Dosimetry

1 **a** The outer casing of the tube is negatively charged, while the central pin is positively charged. When a radiation enters the tube and collides with an electron of one of the gas atoms, it causes ionisation. An electron is removed from its orbit by the collision and speeds towards the central pin, knocking other electrons out of their orbits. This produces an 'avalanche' effect which results in a large group of electrons hitting the central pin to produce a pulse of current which causes a single count on a counter.

    b 1 Bq, 20 Bq, 100 Bq

2 **a** 10,000 decays

    b 200,000 decays

    c $2 \cdot 4 \times 10^6$ decays

    d $7 \cdot 2 \times 10^6$ decays

3 **a** $0 \cdot 2$ Gy

    b $0 \cdot 5$ Gy

    c $0 \cdot 2$ Gy

    d 4 Gy

4 5×10^{-3} Sv, $0 \cdot 015$ Sv, 6×10^{-3} Sv

E **a** grays (Gy)

    b sieverts (Sv)

    c

Absorbed dose (D)	Dose equivalent (H)
$3 \cdot 5$ mGy	$3 \cdot 5$ mSv
$0 \cdot 25$ mGy	$2 \cdot 5$ mSv
15 Gy	15 Sv

    d By increasing the distance from the source

Half-life

1 Half-life is the time taken for the activity of a source to fall to half its initial value.

2 $100 \to 50 \to 25 \to 12 \cdot 5$
Its activity will be $12 \cdot 5$ Bq.

3 $1200 \to 600 \to 300 \to 150 \to 75$
There would be 75 g of molybdenum-101 left.

4 $1000 \to 500 \to 250 \to 125$
3 half-lives = 3 days
The half-life is 1 day.

5

Total Activity (Bq)	250	190	150	120	100	85
Corrected Activity (Bq)	*200	140	*100	70	50	35
Times (minutes	0	5	10	15	20	25

The activity falls from 200 to 100 in 10 minutes. The halflife of the isotope is 10 minutes. You can also see the answer by looking at other values such as $140 \to 70$.

6 $2 \to 1 \to 0 \cdot 5$
The Earth's age is the same as two half-lives of potassium-40. Our estimate of the age of the Earth is $2 \cdot 50 \times 10^9$ years.

E **a** Your graph (if carefully drawn) should give a half-life of 60 minutes. Her conclusion is correct.

    b $8000 \to 4000 \to 2000 \to 1000 \to 500 \to 250 \to 125$
The activity will be 125 Bq.

    c $60 \to 30 \to 15 \to 7 \cdot 5$
After 3 hours, substance X should have an activity of $7 \cdot 5$ kBq. This material cannot be substance X.

Nuclear reactors

1

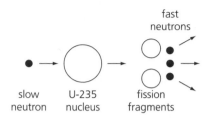

fast
neutrons

slow neutron → U-235 nucleus → fission fragments

2 The bomb reaction is reaction X. The growth of the number of neutrons is very rapid and, happening in a short time, an explosion results.

3 Ist generation: 3^1 neutrons
2nd generation: 3^2 neutrons
5th generation: 3^5 neutrons
$= 3 \times 3 \times 3 \times 3 \times 3 = 243$ neutrons

4 Ideas to include here are:
- radioactive waste products
- long half-lives
- decommissioning of power stations
- terrorism

5 Mention:
- no carbon dioxide emissions
- nuclear fusion in the future
- finite reserves of coal and oil

E a Look again at Q1 on page 46 for help.

b A = control rod
B = uranium fuel rods
C = graphite moderator/reactor core
D = pressure vessel/containment vessel

c **Control rods** absorb neutrons to reduce the amount of fission.
Uranium fuel rods provide the uranium-235 which undergoes the fission reaction to provide heat.
The **core** consists of graphite which slows down the fast neutrons to allow more fission to take place.
The **containment vessel** keeps hot, high-pressure gas in the reactor and helps shield workers from the radiation.

d % efficiency
$= \frac{\text{Useful power output}}{\text{Total power input}} \times 100$
$= \frac{1200 \times 100}{2 \times 1500}$
$= 40\%$

e Advantage: no radioactive decay products
Disadvantage: CO_2 emission